DARWIN'S
THEORY IS FALSE

DARWIN'S THEORY IS FALSE

Three Rules of Scientific Law

Kevin Sparks Berry

Library of Congress Control Number:		2023905938
ISBN:	Hardcover	978-1-6698-7228-3
	Softcover	978-1-6698-7226-9
	eBook	978-1-6698-7227-6

Print information available on the last page.

Rev. date: 03/27/2023

To order additional copies of this book, contact:
Xlibris
844-714-8691
www.Xlibris.com
Orders@Xlibris.com
851740

This book is dedicated to the one true God that
still uses man to communicate with man.

Contents

Prologue

Three Rules of Scientific Law

1. Scientific law is eternal.
2. Scientific law cannot evolve.
3. Scientific law cannot be changed by man.

Each scientific law is the supreme law of all mankind.
Each scientific law cannot be appealed.
Each scientific law is the one truth.

Veritas omnia vincit.
(Translation: Truth conquers all)

Introduction

FW: Darwin's Theory Is False

This book began on Friday, March 11, 2016. I was driving to my last call in Eminence, Missouri. I switched the radio to *Science Friday* on PRI as I cruised down Highway 60. Ira Flatow was talking to a guest about a 99.9 million-year-old fossil of a chameleon which had just been discovered. I thought, *Not 100 million or 99.8 million, but 99.9 million*, and smiled. Then they began to talk about a recent attempt to transplant a uterus. The woman had been born without a uterus. They were going to transplant the uterus so she could have children. She would have to take anti-rejection drugs so her body did not reject the transplanted uterus. After a couple of babies, they would remove the uterus so she could stop taking the anti-rejection drugs. Then he said, "The uterus is not necessary for life." I could not believe my ears! They actually said: "The uterus is not necessary for life." This was so enormously stupid it made me angry. At dinner that evening I told my wife and grandson what they had said but neither was affected like me. So the next morning I began my emails on Darwin's Theory.

I was familiar with the National Center for Science Education's "Project Steve List." The NCSE's Project Steve is a tongue-in-cheek parody of a long-standing creationist tradition of amassing lists of scientists who doubt evolution. There were 1,413 *Steves* that had signed their statement as of March 28, 2017. You can read the statement on the NCSE website. They want the

world to know that tens of thousands of famous scientists believe Darwin's theory is true. What better list of scientists to discuss how the uterus is not necessary for life.

I created my first email and the rest evolved from it. This is vital. Evolution cannot start until something is created. Only then can more follow. This fact the Steves seem to have missed. Evolution involves the gradual change of something from a simple state to something more complex. But to begin there must be something created. Evolution is something from something. Creation is something from nothing.

In my emails, I have noted the parts that evolved from previous emails. The body of the emails remain similar as you will see. Each day's emails are mostly the same but for a slight change. However, on some days, the changes were bigger than others as is true in the evolutionary process. I have also added some clarification on the change if I think necessary. Remember, evolution is from the simple to the more complex.

Emails to Steves

(Project Steve List)

Please ask the NCSE what evolved first, the woman or the uterus?
The science of human reproduction proves Darwin's theory is false,

Sincerely,
Kevin Sparks Berry
Author, *JESUS: THE PROTOTOKOS*

Sent: Sat., 04 June 2016

Subject: FW: Darwin's theory is false

Attn.: Dr. Steven Two (Project Steve List)

The science of human reproduction proves Darwin's theory is false.

**For Darwin's theory to be true, the first female must evolve from the first male, which is impossible.

Sincerely,

Kevin Sparks Berry

Author, *JESUS: THE PROTOTOKOS*

**The first complete female would have to evolve immediately from the first complete male for the first complete male to reproduce. Without the first complete female, the first complete male would go extinct.

Sent: Mon., 06 Jun 2016

Subject: FW: Darwin's theory is false

Attn.: Dr. Steven Three (Project Steve List)

The science of human reproduction proves Darwin's theory is false.

For Darwin's theory to be true, the first female must evolve from the first male, which is not scientifically possible.

**Biological science proves Darwin's theory is false.

Sincerely,

Kevin Sparks Berry

Author, *JESUS: THE PROTOTOKOS*

**Every ninth-grade student is taught in biology class there must be a complete female for a complete male to reproduce.

Sent: Tues., 07 Jun 2016

Subject: FW: Darwin's theory is false.

Attn.: Dr. Stephen James Four (Project Steve List)

The science of human reproduction proves Darwin's theory is false.

For Darwin's theory to be true, the first female must evolve from the first male, which is not scientifically possible.

Therefore, Darwin's theory is false.

Sincerely,

Kevin Sparks Berry

Author, *JESUS: THE PROTOTOKOS*

Sent: Fri., 24 Jun 2016

Subject: FW Darwin's theory is false.

Attn.: Dr. Stephen J. Five (Project Steve List)

The science of human reproduction proves Darwin's theory is false.

For Darwin's theory to be true the first female must evolve from the first male, which is not scientifically possible.

Therefore, Darwin's theory is false.

Sincerely,

Kevin Sparks Berry

Author, *JESUS: THE PROTOTOKOS*

**PS: Man has free will to believe Darwin's theory or not.

**Even though Darwin's theory is false, man can believe it or not.

Sent: Sat., 25 Jun 2016
Subject: FW: Darwin's theory is false.
Attn.: Dr. Steven Six (Project Steve List)

The science of human reproduction proves Darwin's theory is false.

For Darwin's theory to be true, the first female must evolve from the first male, which is not scientifically possible.

Therefore, Darwin's theory is false.

Sincerely,
Kevin Sparks Berry
Author, *JESUS: THE PROTOTOKOS*

**PS: There is no reason for the first female to begin evolving until the first male evolves, and then she is needed immediately.

**If monosexual reproduction evolved first, there would be no necessity for bisexual reproduction. Monosexual reproduction is more efficient. If a first male did evolve, he would not have the intelligence to know he needed a female to reproduce.

Sent: 26 Jun 2016
Subject: FW: Darwin's theory is false.
Attn.: Dr. Steven M. Seven (Project Steve List)

The science of human reproduction proves Darwin's theory is false.

For Darwin's theory to be true, the first female must evolve from the first male, which is not scientifically possible.

Therefore, Darwin's theory is false.

Sincerely,
Kevin Sparks Berry
Author, *JESUS: THE PROTOTOKOS*

**PS: Evolution is completely random. Evolution has zero intelligence. Under evolution, after the first male evolves, there is no reason for a female to evolve.

**Evolution is just a process which is random. It does not have a plan for the future. Evolution has zero intelligence on its own with which to create anything. Evolution would not know a female is needed for the male to reproduce.

Sent: Tue., 28 Jun 2016
Subject: FW: Darwin's theory is false.
Attn.: Dr. Steven W. Eight (Project Steve List)

The science of human reproduction proves Darwin's theory is false.

For Darwin's theory to be true, the first female must evolve from the first male, which is not scientifically possible.

Therefore, Darwin's theory is false.

Sincerely,

Kevin Sparks Berry

Author, *JESUS: THE PROTOTOKOS*

**PS: Evolution is completely random. Evolution has zero intelligence. Evolution does not create intelligence.

**Evolution cannot create anything, especially intelligence.

Sent: Fri., 01 Jul 2016

Subject: FW: Darwin's theory is false.

Attn.: Dr. Stephen H. Nine (Project Steve List)

The science of human reproduction proves Darwin's theory is false.

For Darwin's theory to be true, the first female must evolve from the first male, which is not scientifically possible.

Therefore, Darwin's theory is false.

Sincerely,

Kevin Sparks Berry

Author, *JESUS: THE PROTOTOKOS*

PS

**Truth is a beautiful thing, and there is only one.

**Truth is beautiful. Each scientific law is the one truth. Each scientific law is singular.

Sent: Sun., 17 Jul 2016
Subject: FW: Darwin's theory is false.
Attn.: Dr. Steven M. Ten (Project Steve List)

The science of human reproduction proves Darwin's theory is false.

For Darwin's theory to be true, the first female must evolve from the first male, which is not scientifically possible.

Therefore, Darwin's theory is false.

Sincerely,

Kevin Sparks Berry

Author, *JESUS: THE PROTOTOKOS*

PS Evolution is completely random. Evolution has zero intelligence. Evolution does not create intelligence.

**Something cannot evolve from nothing

**There is nothing in eternity's past. Eternity's past is completely void. Something cannot evolve from nothing.

Sent: Mon., 18 Jul 2016

Subject: FW: Darwin's theory is false.

Attn.: Dr. Stephen Eleven (Project Steve List)

The science of human reproduction proves Darwin's theory is false.

For Darwin's theory to be true, the first female must evolve from the first male, which is not scientifically possible.

Therefore, Darwin's theory is false.

Sincerely,

Kevin Sparks Berry

Author, *JESUS: THE PROTOTOKOS*

**PS: SOMETHING CANNOT EVOLVE FROM NOTHING.

**I put this in all caps to emphasize the fact!

Sent: Fri., 22 Jul 2016
Subject: FW: Darwin's theory is false.
Attn.: Dr. Stefanie D. Z. Twelve (Project Steve List)

The science of human reproduction proves Darwin's theory is false.

For Darwin's theory to be true, the first female must evolve from the first male, which is not scientifically possible.

Therefore, Darwin's theory is false.

Sincerely,
Kevin Sparks Berry
Author, *JESUS: THE PROTOTOKOS*

**PS: Science cannot prove evolution creates intelligence. Science cannot prove something evolves from nothing. Therefore, science cannot prove Darwin's theory is true.

**Science has been trying to prove evolution creates intelligence but cannot. It is impossible for science to prove something can evolve from nothing. Something must first be created for something to evolve from it.

Sent: Wed., 27 Jul 2016

Subject: FW: Darwin's theory is false

Attn.: Dr. Stephen J. Thirteen (Project Steve List)

The science of human reproduction proves Darwin's theory is false.

For Darwin's theory to be true, the first female must evolve from the first male, which is not scientifically possible.

Therefore, Darwin's theory is false.

Sincerely,

Kevin Sparks Berry

Author, *JESUS: THE PROTOTOKOS*

**PS: Science will never prove evolution creates intelligence. Science will never prove something evolves from nothing. Science will never prove there was a big bang.

** For the Big Bang theory to be true, there must be something to make a big bang. There is nothing in eternity's past.

Sent: Fri., 29 Jul 2016

Subject: FW: Darwin's theory is false.

Attn.: Dr. Steven Fourteen (Project Steve List)

The science of human reproduction proves Darwin's theory is false.

For Darwin's theory to be true, the first female must evolve from the first male, which is not scientifically possible.

Therefore, Darwin's theory is false.

Sincerely,

Kevin Sparks Berry

Author, *JESUS: THE PROTOTOKOS*

**PS: Science will never prove evolution creates intelligence. Science will never prove something evolves from nothing. Therefore, the Big Bang theory is dead as well.

Sent: Sun., 31 Jul 2016

Subject: FW: Darwin's theory is false

Attn.: Dr. Stephen D. Fifteen (Project Steve List)

The science of human reproduction proves Darwin's theory is false.

For Darwin's theory to be true, the first female must evolve from the first male, which is not scientifically possible.

Therefore, Darwin's theory is false.

Sincerely,

Kevin Sparks Berry

Author, *JESUS: THE PROTOTOKOS*

**PS: Science will never prove evolution creates intelligence. Science will never prove something evolves from nothing. Science will never prove quantum fluctuations evolved from nothing.

Sent: Fri., 05 Aug 2016

Subject: FW: Darwin's theory is false

Attn.: Dr. Steven N. Sixteen (Project Steve List)

    **The science of evolution proves Darwin's theory is false.

    **For Darwin's theory to be true, it must start at the beginning of eternity past. At the beginning of eternity's past, there was no space or time.

    Something cannot evolve from nothing; therefore, Darwin's theory is false.

Sincerely,

Kevin Sparks Berry

Author, *JESUS: THE PROTOTOKOS*

PS: Science will never prove evolution created intelligence. Science will never prove something evolved from nothing. Science will never prove quantum fluctuations evolved from nothing.

Sent: Fri., 12 Aug 2016
Subject: FW: Darwin's theory is false
Attn.: Dr. Steven Seventeen (Project Steve List)

The science of evolution proves Darwin's theory is false.

There are three segments of time: (1) the time of eternity's past, (2) the time of eternity's present, and (3) the time of eternity's future.

For Darwin's theory to be true, it must start in the beginning which is segment 1.

There is nothing in segment 1. There is something in segment 2.

Something cannot evolve from nothing; therefore, Darwin's theory is false.

Sincerely,

Kevin Sparks Berry

Author, *JESUS: THE PROTOTOKOS*

PS: Science will never prove evolution created intelligence. Science will never prove something evolves from nothing. Science will never prove quantum fluctuations evolved from nothing.

Sent: Tue., 23 2016

Subject: FW: Darwin's theory is false

Attn.: Dr. Stephanie Malia Eighteen (Project Steve List)

    **The science of DNA proves Darwin's theory is false.

    **For Darwin's theory to be true, every man must have monkey DNA in his ancestry.

    **Please show me one.

Sincerely,

Kevin Sparks Berry

Author, *JESUS: THE PROTOTOKOS*

PS: I am from Missouri, the Show Me State.

Sent: Sun., 04 Sep 2016

Subject: FW: Darwin's Theory is False.

Attn.: Dr. Stephen M. Nineteen (Project Steve List)

The science of DNA proves Darwin's theory is false.

**The science of DNA tells us who we are related to in the past.

**The science of DNA tells us we are not related to chimps in the past.

Therefore, Darwin's theory is false.

Sincerely,

Kevin Sparks Berry

Author, *JESUS: THE PROTOTOKOS*

PS: Darwin did not have the knowledge of DNA in 1859 that we do today.

**Known rates of gene mutation are limited to a few thousand years.

Sent: 23 Jan 2017

Subject: FW: Darwin's theory is false.

Attn.: The Royal Society

P1. The science of DNA tells us who we descended from.

P2. The science of DNA tells us we only descended from Homo sapiens.

P3. The science of DNA has never told anyone they descended from non-Homo sapiens.

C: Therefore, Darwin's theory is false.

My theory: When ancestry.com gets back 6,000 years, there will be no more ancestor DNA.

Sincerely,

Kevin Sparks Berry

Author, *JESUS: THE PROTOTOKOS*

PS: The Royal Society is a fellowship of many of the world's most eminent scientists and is the oldest scientific academy in continuous existence.

Summary

1. Science of human reproduction:

For Darwin's theory to be true, the first female must evolve from the first male, which is not scientifically possible.

Therefore, Darwin's theory is false.

2. Science of evolution:

For Darwin's theory to be true, evolution must begin in segment 1, which is not scientifically possible. Something cannot evolve from nothing.

Therefore, Darwin's theory is false.

3. Science of DNA:

The science of DNA tells us who we descended from. The science of DNA tells us we only descended from Homo sapiens. The science of DNA has never told anyone they descended from non-Homo sapiens.

Therefore, Darwin's theory is false.

Epilogue

Farmington City Council
10/24/2022

Darwin's Theory is False
By Kevin Sparks Berry
Three Rules of Scientific Law

1. Scientific Law is eternal
2. Scientific Law cannot evolve
3. Scientific Law cannot be changed by man

Begin Time		End Time
Eternity Past	Eternity Present	Eternity Future
ϕ Complete Void	$\leftarrow$ Law of Universal Gravitation $\rightarrow$ $(F = G\dfrac{m_1 m_2}{r^2})$	Time is not necessary because there is no end.
	$\leftarrow$ Law of Conservation of Mass $\rightarrow$ $(\dfrac{\partial p}{\partial t} + \nabla \Box (pv) = 0)$	
There can be no time because there is no beginning	$\leftarrow$ Law of Conservation of Energy $\rightarrow$ $(K_1 + U_1 = K_2 + U_2)$	
	$\leftarrow$ Law of Cosmic Expansion $\rightarrow$ $(V = H_0 D)$	
	$\leftarrow$ Law of Planetary Motion $\rightarrow$ $(P^2 = a^3)$	
	$\leftarrow$ Law of Eternal Life $\rightarrow$ (He that believeth and is baptized shall be saved.)	

Something cannot evolve from nothing. Therefore, **Darwin's Theory is False.**

Final Thought

Plus ultra.
(Translation: There is more beyond.)

This was given to me by Pastor John L. Carnett at the Graveside Service of Earl Hesterberg, my beloved father-in-law, on November 9, 2022.

www.ingramcontent.com/pod-product-compliance
Lightning Source LLC
Chambersburg PA
CBHW021511210526
45463CB00002B/978

* 9 7 8 1 6 6 9 8 7 2 2 6 9 *